SAFETY AT WORK

SAFETY AT WORK

A Gower Health and Safety Workbook

Graham Roberts-Phelps

Gower

Published by
Gower Publishing Limited
Gower House
Croft Road
Aldershot
Hampshire GU11 3HR
England

Gower
Old Post Road
Brookfield
Vermont 05036
USA

British Library Cataloguing in Publication Data
Roberts-Phelps, Graham
Safety at work. – (A Gower health and safety workbook)
1.Industrial safety – Great Britain 2.Industrial safety – Law and legislation – Great Britain
I.Title
363.1′1′0941

ISBN 0 566 08067 2

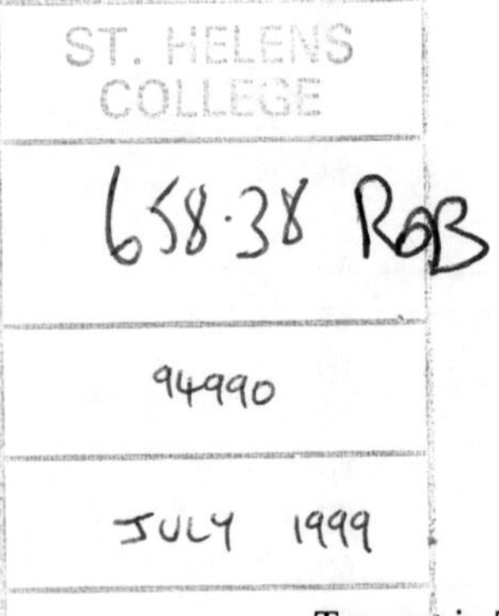

Typeset in Times by Wearset, Boldon, Tyne and Wear and printed in Great Britain by print in black, Midsomer Norton.

Contents

Chapter 1
Introduction

This first chapter acts as a record of your progress through the workbook and provides a place to summarize your notes and ideas on applying or implementing any of the points covered.

PERSONAL DETAILS

Name:	
Position:	
Location:	
Date started:	Date completed:

Chapter title	Signed	Date
1. Introduction		
2. Understanding safety legislation		
3. How to improve your safety rating		
4. Attitude makes the difference		
5. Learning review		
Demonstration of safety in the workplace		
Steps taken to reduce risks and hazards		

Safety review dates	Assessed by	Date
1 month	________________	________________
2 months	________________	________________
3 months	________________	________________
6 months	________________	________________

HOW TO USE THIS SELF-STUDY WORKBOOK

Overview

This self-study workbook is designed to be either one, or a combination, of the following:

- a self-study workbook to be completed during working hours in the student's normal place of work, with a review by a trainer, manager or safety officer at a later date

- a training programme workbook that can be either fully or partly completed during a training event or events, with the uncompleted sections finished in the student's normal working hours away from the training room.

It contains six self-contained chapters which should each take about 20 minutes to complete, with the final section, 'Learning Review', taking slightly longer due to the testing and validation instruments.

It is essential that students discuss their notes and answers from all sections with a supervisor, trainer or coach.

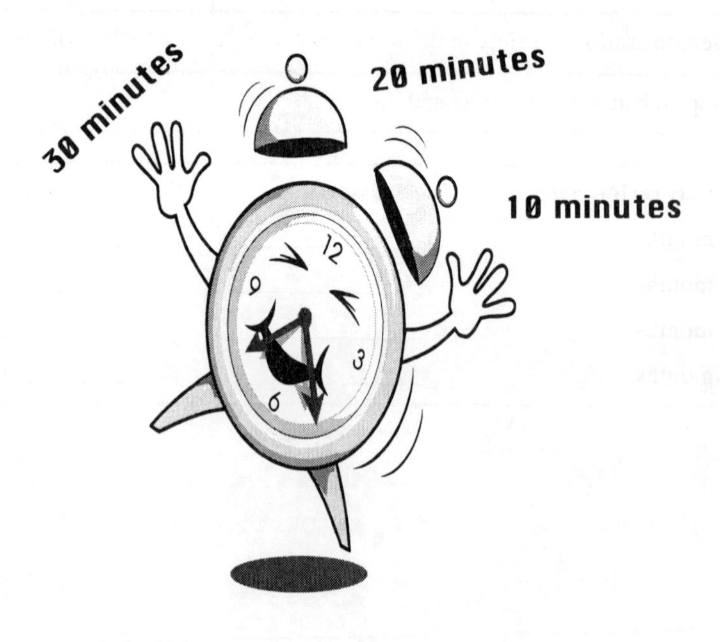

NOTES FOR TRAINERS AND MANAGERS

For use in a training session

If you are using the workbook in a training event you might choose to send the manual to students in advance of their attendance, asking them to complete the Introduction (Chapter 1). Other exercises can then be utilized as required during the programme.

For use as an open-learning or self-study tool

Make sure that you have read the workbook properly yourself and know what the answers are. Anticipate any areas where students may require further support or clarification.

Comprehension testing

Each section features one or more summary exercises to aid understanding and test retention. The final chapter, 'Learning Review', contains a set of tests, case studies and exercises that test application and knowledge. Suggested answers to these are given in the Appendix.

If you are sending the workbook out to trainees, it is advisable to send an accompanying note reproducing, or drawing attention to, the points contained in the section 'Notes for Students'. Also, be sure to set a time deadline for completing the workbook, perhaps setting a review date in advance.

The tests contained in the learning review can be marked and scored as a percentage if required. You might choose to set a 'pass' or 'fail' standard for completion of the workbook, with certification for all those attaining a suitable standard. Trainees who do not reach the required grade on first completion might then be further coached and have points discussed on an individual basis.

NOTES FOR STUDENTS

This self-study workbook is designed to help you better understand and apply the topic of safe manual handling. It may be used either as part of a training programme, or for self-study at your normal place of work, or as a combination of both.

Here are some guidelines on how to gain the most from this workbook.

- Find 20 minutes during which you will not be disturbed.
- Read, complete and review one chapter at a time.
- Do not rush any chapter or exercise – time taken now will pay dividends later.
- Complete each written exercise as fully as you can.
- Make notes of questions or points that come to mind when reading through the sections.
- Discuss anything that you do not understand with your manager, safety officer or work colleagues.

The final chapter, 'Learning Review', is a scored test that may carry a pass or fail mark.

At regular intervals throughout the workbook there are exercises to complete and opportunities to make notes on any topics or points that you feel are particularly important or relevant to you. These are marked as:

Notes

LEARNING DIARY

Personal Learning Diary

Name: ______________________________

Job Title: ______________________________

Company: ______________________________

Date: ______________________________

The value of the training programme will be greatly enhanced if you complete and review the following Learning Diary before, during and after reviewing and reading the workbook.

Learning Objectives

At the start or before completing the workbook, please take time to consider what you would like to learn or be able to do better as a result of the training process. Please be as specific as possible, relating points directly to the requirements of your job or work situation. If possible, please involve your manager, supervisor or team leader in agreeing these objectives.

Record these objectives below

1.

2.

3.

4.

5.

6.

Please complete before continuing

Learning Log

During the training programme there will be many useful ideas and learning points that you will want to apply in the workplace.

Key ideas/learning points	How I will apply these at work

Please complete before continuing

LEARNING APPLICATION

As you complete each chapter, please consider and identify the specific opportunities for applying the skills, knowledge, behaviours and attitudes and record these below.

Action planned, with dates	Review/comments

Remember, it may take time and practice to achieve new results.
Review these goals periodically and discuss with your manager.

PLEASE COMPLETE BEFORE CONTINUING

HOW TO GET THE BEST RESULTS FROM THIS WORKBOOK

The format of this workbook is interactive; it requires you to complete various written exercises. This aids both learning retention and comprehension and, most importantly, acts as a permanent record of completion and learning. It is therefore essential that you **complete all exercises, assignments and questions**.

In order to gain the maximum value and benefit from the time that you invest in completing this workbook, use the following guidelines.

Pace yourself

You might choose to work through the whole workbook in one session or, alternatively, you might find it easier to take one chapter at a time. This is the recommended approach. If you are using this workbook as part of a live training programme, then make time to follow through any unfinished exercises or topics afterwards.

Share your own opinions and experience

We all have a different view of the world, and we all have different backgrounds and experiences. As you complete the workbook it is essential that you relate learning points directly to your own situation, beliefs and work environment.

Please complete the exercises using relevant examples that are personal and specific to you.

Keep an open mind

Some of the material you will be covering may be simple common sense, and some of it will be familiar to you. Other ideas may not be so familiar, so it pays to keep an open mind, as most learning involves some form of change. This may take the form of changing your ideas, changing an attitude, changing your perception of what is true, or changing your behaviours and the way you do things.

When we experience change, in almost anything, our first automatic reaction is resistance, but this is not usually the most useful response. Remember, safety is something we have been aware of for a long time, and consider (or fail to consider, as the case may be!) every day. As a result, we follow procedures without thinking – on auto-pilot as it were. This often means that we have a number of bad habits of which we are unaware.

Example of change:

Sign your name here as you would normally do:

Now hold the pen or pencil in the ***opposite*** *hand to that which you normally use and sign your name again:*

Apart from noting how difficult this might have been, consider also how 'strange' and uncomfortable this seemed. You could easily learn to sign your name with either hand, usually far more quickly than you might think. However the resistance to change may take longer to overcome.

Make ***Notes***

Making notes not only gives you information to refer to later, perhaps while reviewing the workbook, but it also aids memory. Many people find that making notes actually helps them to remember things more accurately and for longer. So, as you come across points that are particularly useful or of particular interest, please take a couple of moments to write these down, underline them or make comments in the margin or spaces provided.

Review with others

In particular, ask questions and discuss your answers and thoughts with your colleagues and fellow managers, especially points which you are not sure of, points which you are not quite clear on, and perhaps points about which you would like to understand more.

Before you start any of the main chapters, please complete the following learning assignments.

Learning Objectives

It is often said that if you do not know where you are going, any road will get you there. To put it another way, it is difficult to hit the target you cannot see. To gain the most benefit from this workbook, it is best to have some objectives.

Overall objectives

- **Improvements.** We don't have to be ill to improve our fitness. Improvement is always possible.
- **Skills.** Learn new skills, tips and techniques.
- **Knowledge.** Gain a better understanding of safety issues.
- **Attitudes.** Change the way we think about safety issues.
- **Changes.** Change specific attitudes on behaviours and our safety procedures and practice.
- **Ideas.** Share ideas.

Here are some areas in which you can apply your overall objectives.

1. Hazards and risks

The first objective is to be able to identify safety hazards and risks. These may exist all around us and may not be readily identifiable as such – for example, the ordinary moving of boxes or small items, using a kettle or hand drill, cleaning and so on.

2. Prevention

Prevention is always better than cure, and part of this workbook will deal with knowing how to prevent accidents and injuries in the first place. Injuries are nearly always painful both in human and business terms. As well as accidents that cause us or others harm, there are many more accidents that cause damage and cost money to put right.

3. Understanding your safety responsibilities

Health and safety is everybody's responsibility, and safety is a full-time job. As you complete this workbook you will be looking at how it affects you personally and the role that you can play, not only for your own safety but also for the safety of others around you.

4. Identifying ways to make your workplace safer

A workbook like this also gives us the opportunity to put ideas together on how we can improve the health and safety environment of our workplace. We do not have to have safety problems in order to improve safety, any more than we have to be ill to become fitter.

An improvement in working conditions does not have to cost much or be very complicated. Simply moving a filing cabinet to a more convenient location often represents a quantum leap towards working safely.

Make a note here of any personal objectives that you may have.

Notes

Reasons to Learn

The various studies that have been undertaken on how and why people learn and why some people learn more quickly than others have discovered that motivation plays a significant role in our ability to learn.

When answering both the questions set below, please consider not only your own personal situation but those of the company, the organization, your work colleagues and, possibly, your customers.

1. *What **difficulties** or **disadvantages** derive when people are not very aware of good safety methods and practices?*

2. *What **benefits** or **advantages** derive when people are more health and safety conscious and skilled?*

Consequences of poor safety: Consider not only business costs, but costs to you personally. These may include lost overtime and bonuses, cost of medical prescriptions, missed work opportunities and disruptions to your social life and hobbies.

PLEASE COMPLETE BEFORE CONTINUING

Opinion Poll

Consider the following statements, first marking each with your level of agreement, and then making some supporting comments regarding these views.

5 = Strongly agree; 4 = Agree; 3 = Neither agree nor disagree; 2 = Disagree; 1 = Strongly disagree.

1. Every accident or injury can be prevented or avoided.

Circle one response: 5 4 3 2 1

Comments:

2. Every accident or work-related injury or discomfort is caused by human error in some way.

Circle one response: 5 4 3 2 1

Comments:

3. You cannot motivate people to be safer; you can only enforce rules and penalties.

Circle one response: 5 4 3 2 1

Comments:

4. Left to their own devices, people and organizations will take unnecessary risks and cut corners.

Circle one response: 5 4 3 2 1

Comments:

PLEASE COMPLETE BEFORE CONTINUING

Opinion Poll: Review

1. Every accident or injury can be prevented or avoided.

This is largely held to be true. Research shows that nearly all accidents are a result of a cause and effect relationship. If you identify the causes, you can change the effects.

2. Every accident or work-related injury or discomfort is caused by human error in some way.

As a computer programmer once remarked, 'There is no such thing as "computer error", only incorrect user input'. 'Accidents' are caused by people and their behaviours, not by machines, chemicals or inanimate objects.

3. You cannot motivate people to be safer; you can only enforce rules and penalties.

Hopefully, people will work safely and consider their own welfare and that of others without legal or management interference, although statistics do not prove this to be the case. In countries without enforced legislation, people are made to endure terrible work environments with little or no regard for safety. Consider how many of us wear a seat belt today compared with the number who did so before it became law.

4. Left to their own devices people and organizations will take unnecessary risks and cut corners.

Accident investigators and HSE inspectors have thousands of examples which prove this statement to be true. You cannot have a quality company that does not consider the health and safety of its staff and customers as the highest priority.

Chapter 2
Understanding Safety Legislation

This chapter examines the current regulations and standards of safe work practice.

Before starting this chapter, please take a few moments to make a note of any ideas or actions in the learning diary and log in Chapter 1.

The biggest risk is not taking Health and Safety seriously.

HEALTH AND SAFETY LEGISLATION

Common law requires that an employer must take reasonable care to protect his employees from risk of foreseeable injury, disease or death at work. In the nineteenth and twentieth centuries employers argued with reasonable success against this duty. It was not until 1938 that the House of Lords identified, in general terms, the duties of employers as common law.

Whilst it is obviously good common sense to work safely, minimizing the chances of accidents, it is also a point of law. There are two main kinds of Health and Safety law. Some is very specific about what you must do, and some is much more general, applying to almost every business. In this short summary we will be looking at the key legislation that affects your work and translating it into practical measures by which we must all abide.

Never underestimate the consequences of breaking Health and Safety legislation. We can all remember terrible accidents such as the Zeebrugge disaster or the Clapham rail accident. However, there are many hundreds of thousands of accidents each year that do not make the headlines but still ruin lives.

Legally, accidents like these can cost companies and individuals thousands and sometimes hundreds of thousands of pounds.

> ***Example:*** *A meat processing plant was burnt to the ground by a fault caused during work on a piece of machinery. The fire alarm and fire sprinklers failed to operate properly – both legal requirements. Fortunately, nobody was hurt.*
>
> *Any fines imposed by the courts would be insignificant compared to the loss of business, customer contracts and the effect on 100 employees who lost their jobs for a year while the factory was rebuilt.*

Some risks are very obvious; others less so. The purpose of legislation is that all possible steps are taken to eliminate hazards, reduce risks, and inform and implement safe systems of work.

25% of all fatal accidents, and many more serious injuries, are caused because safe systems of work are not provided for, or even ignored.

Health and Safety law: what does it mean to you?

Safety regulations or legislation create very real obligations not only on companies and organizations, but also on the directors, managers and individual employees. Whilst some are over 20 years old, many are much more recent and it is important that we are fully aware of the consequences and the requirements of each set of regulations. As the law says, ignorance is no defence.

If there is an accident, Health and Safety laws are interpreted so that the company or the organization, its managers and directors, have to prove that it was not at fault. In other words, it is assumed that the organization has not met its Health and Safety obligations, unless it can prove otherwise. Therefore, failure to comply in a way that demonstrates and proves that legislation has been adhered to can easily lead to prosecution, resulting in fines and, in some cases, imprisonment.

Some legislation that may affect you

- Health and Safety at Work Act 1974
- Electricity at Work Regulations 1989
- COSHH Regulations 1994
- Manual Handling Operations Regulations 1992
- Noise at Work Regulations 1989
- Fire Precautions Act 1971
- Display Screen Equipment Regulations 1992
- Workplace (Health, Safety and Welfare) Regulations 1992
- Management of Health and Safety at Work Regulations 1992
- Provision and Use of Work Equipment Regulations 1992
- Safety Signs and Signals Regulations 1996
- Consultation with Employees Regulations 1996

In law, ignorance is no defence.

Make a note of any points from this section that concern you.

Notes

HEALTH AND SAFETY AT WORK ACT 1974

This is the main law that covers **everyone** at work and **all work premises**.

It simply means making sure that people work safely, are safe and that their welfare is not put at risk.

Enacted in 1974, this piece of legislation was brought in to replace and update much of the old Health and Safety law that was contained in the Factories Act 1961 and the Offices, Shops and Railway Premises Act 1963. These two laws were rapidly becoming out-of-date with the advent of modern working practices, technology and equipment.

Under the Health and Safety at Work Act (HASAWA) a company has to ensure the health and safety of all its employees. Individuals have to ensure the health and safety of themselves and others around them who may be affected by what they do, or fail to do. This includes contractors, as well as customers or, indeed, anyone who may come into contact with the organization.

Safety studies have highlighted that small firms (those with fewer than 50 employees) have a higher incidence of accidents than large organizations. The figures may be even worse because of the large number of accidents that go unreported by small companies. Ignorance, poor standards and contempt for basic safety standards have been highlighted as the key contributing factors.

The Act applies to all work activities and premises, and everyone at work has responsibilities under it – including the self-employed. Here are some key points raised by the Act.

1. Safeguards

Employers are required to implement reasonable safeguards to ensure safe working practice at all times. This means taking every practical step to remove hazards and reduce or eliminate risks. The law interprets this as taking every possible precaution, and cost is not considered as an excuse for failure to do this.

2. Written policy

All organizations employing five or more people must have a written and up-to-date health and safety policy. In addition, they must carry out written risk assessments as part of the implementation of their safety policy and also display a current certificate as required by the Employer's Liability (Compulsory Insurance) Act 1969.

3. Training and information

Following on from this, all staff must be fully trained, equipped and informed of not only the company's safety policy and procedures but also of the skills and knowledge necessary to carry out their normal work duties. This, of course, means displaying Health and Safety regulations and safety signs, as well as formally training and directing staff on all aspects of health, safety, hazards and risks.

4. Reasonable care

The legislation does not just cover employers and organizations; there are definite requirements placed on employees. All employees must take reasonable care not only to protect themselves, but also their colleagues. They are also required to comply with all health and safety policy regulations and procedures in full and to cooperate fully with health and safety representatives and officers in their job of implementing Health and Safety policies. Failure to do so is in breach of the Act.

HSE Inspectors can visit without notice and have right of entry. They have the power to stop your work, close premises and even prosecute.

---

5. Safe systems of work

Employers must also ensure what are known as 'safe systems of work', which means creating an environment that is conducive to health and safety. This can be as basic as making sure that buildings are in good repair, that proper heat and ventilation are provided, and that the workplace is clean and hygienic to work in. However, it may also mean having clear procedures and checklists to make sure that safety is implemented. In some cases, a permit to work may be required in order to carry out certain jobs. A 'safe system of work' should also document what to do in the event of accidents and emergencies.

Make a note of any points from this section that concern you.

Notes

THERE IS NO SUCH THING AS AN ACCIDENT

Workplace accidents are caused by people. More accurately, they are caused by what they do or don't do.

Equipment and machinery may sometimes break down and incidents may occur which cause accidents, but they are nearly always traceable to some degree of human error, negligence or ignorance.

Workplace accidents happen to ordinary people. Although we may like to think that accidents only happen to other people, or believe that we are somehow cleverer, better, luckier or more organized than other people, in reality, an accident can happen to any of us at any time. It could happen to you or me; it could happen today or tomorrow, next month or next year.

How aware are you of safety in your day-to-day work?

> ***Study assignment:***
>
> *Look around your normal workplace, and the room in which you are working now, and identify as many potential hazards as you can. Find at least five.*
>
> *1.*
>
> *2.*
>
> *3.*
>
> *4.*
>
> *5.*

The dictionary defines an accident as 'an unforeseen event' or 'a misfortune or mishap, especially causing injury or death'. However, the safety statistics tell us that we can predict accidents because we know what the causes are and, if we see the causes occurring, we can be sure that there is an accident waiting to happen somewhere along the line.

Accidents do not happen by themselves; they are caused by people like you or me not taking safety seriously.

Make some notes on recent accidents, injuries or illnesses which you or your work colleagues have suffered.

Notes

Accident Statistics

How many people do you think suffer disabling injuries every year whilst at work?

(a) 10 000 or less b) 100 000 c) 1 000 000 or more

How many people do you think are killed at work every year?

a) 10 or less b) 100 c) 500 or more

How many working days do you think are lost every year in the UK because of accidents, sickness or injury?

a) 1 million b) 5 million c) 10 million or more

What do you think are the three most common accidents at work?

1.
2.
3.

Please complete before continuing

ACCIDENT STATISTICS

For an industrialized nation like ours, with a reasonably good record of health and safety awareness, the statistics are quite surprising. Every year over 30 million working days are lost because of work-related accidents, sickness or injury.

A casual attitude can often result in a casualty!

There are literally hundreds of thousands of workplace accidents every year, and, at any one time, several million people are suffering ill-health, either caused or made worse by work conditions. Furthermore, on average, every working day there are at least two fatal injuries in the workplace. This means that tonight, somebody, somewhere, will not be going home.

Accidents can happen to any of us at any time. They are not a rarity. Fortunately, they are also not that common, but we do need to make sure that we do not become another statistic.

Workplace accidents are more common than you might think

Here are some more painful statistics illustrating the most common types of accident and their associated causes. The percentages below derive from the Health and Safety Commission's *Annual Statistics Report* and refer to the most commonly occurring accidents that have been reported. (There are, of course, many more – possibly a much larger number – that go unreported.)

For employees:

1. **Slips, trips and falls (on the same level)** **35%**
2. **Falls from height** **21%**
3. **Injuries from moving, falling or flying objects** **12%**

For the self-employed:

1. **Falls from height** **45%**
2. **Slips, trips and falls (on the same level)** **15%**
3. **Injuries from moving, falling or flying objects** **14%**

These statistics are for the most recent period available. However, the HSC comment that 'whilst most other accidents stayed relatively unchanged, slip, trip or fall accidents have increased from 26% to 35% for employees for the period 1986 to 1996'. The above figures do not take into account illness or sickness that may be caused by repetitive strain or cumulative injuries.

So, in summary, there are 1.6 million accidents at work each year; 2.2 million people are currently suffering ill-health caused, or made worse, by work conditions; 30 million working days per year are lost; and every year about 500 people are killed at work and several thousand more are permanently disabled through work-related accidents or injury.

The most common workplace accidents

- **Straining the body**
  – twisting, reaching or stretching
- **Moving or falling objects**
  – most common damage to head, fingers, feet and eyes
- **Slips, trips and falls**
  – from minor grazes to a broken neck
- **Getting caught in a machine**
  – belts, pulleys, slicers, grinders and so on
- **Injuries caused by hazardous chemicals**
  – inhalation or direct exposure
- **Hearing loss or damage**
  – loud noise can destroy or damage hearing
- **Electric shock**
  – almost **any** electric tool or appliance can kill
- **Eye damage**
  – flying objects, splashing liquids, intense heat or light

A moment's carelessness – a lifetime's regret.

Please make a note of any points from this section that concern you.

Notes

WORKPLACE SAFETY AWARENESS

Many people, when first discussing the topic of workplace safety, ask the question, 'But our workplaces are safe, aren't they?'. The answer, of course, is that they can be, but only if we make safety a priority. Your workplace is as safe, or as dangerous and hazardous, as you choose to make it. However, workplace 'accidents' may not be accidents at all.

Many workplace accidents are cumulative and take years to take effect. For instance, RSI, or repetitive strain injury, can be the result of years of neglect or poor practice in doing repetitive tasks – often involving very small repetitive movements, as in using a keyboard or operating machinery – in an intense or prolonged fashion. The effects of these injuries are sometimes very far-reaching.

Good housekeeping

Much of good safety is really just good common sense and as we walk around our workplaces on a daily basis we must keep in mind what could generally be summarized as good housekeeping. In 1992 a new series of Regulations, the Workplace (Health, Safety and Welfare) Regulations 1992, were introduced to update the ageing Offices, Shops and Railway Premises Act 1963 and the Factories Act 1961. These new Regulations take account of new working conditions, new working styles and new technology in the work environment. However, the basic fundamental principles remain the same – think safety; act safety; be safe.

Here are seven keys to good safety awareness.

1. Walk areas

Walk areas must be kept clear and tidy. This does not just apply to emergency exits, fire exits or areas through which customers may be walking. All walk areas must be kept clear and free of debris, rubbish, boxes and other obstructions.

Think Safety!

- **But we work safely, don't we?**
- **... only if we make safety a priority.**
- **Many 'accidents' are cumulative and take years to take effect.**
- **A moment's carelessness – a lifetime of regret!**

Being safe simply means knowing what to do (and what NOT to do) and then DOING what you know.

2. Drawers

Leaving drawers open can cause a very annoying and pointless accident – particularly as they are so simple to close!

3. Chemicals

Chemicals must be stored and labelled correctly, whether these are correcting fluids, floor cleaners, polishes, abrasives or specialized chemicals used in our work. Failure to do so is breaking the law.

4. Ventilation and heating

The organization for which you work must provide a reasonable working temperature in all workrooms, local heating or cooling systems where a comfortable temperature cannot be maintained and good ventilation. Draughts and heating systems giving off dangerous or offensive levels of fumes must be avoided. In addition, workrooms should be spacious enough to work comfortably in, and arrangements should be made to protect non-smokers from discomfort caused by tobacco smoke.

5. First aid

Every workplace environment should have first aid equipment, a trained first-aider and detailed procedures to follow in the event of personal accident or injury. There should also be an accident book and a record kept of all uses of first aid equipment.

6. Noise and hygiene

These must be controlled. Clean and well ventilated toilets with wash basins, hot and cold running water and drinking water must be provided. Noise should either be eliminated or kept to an absolute minimum and, where noisy environments cannot be eliminated, protective equipment must be issued and worn.

7. Safety signs

These must be provided at all appropriate points and for all valid reasons. They must be in good condition and clearly displayed. People should also be aware of their meaning, either through explanation or training. Whilst many are very self-explanatory, others are slightly more complicated and everyone will need to understand what these mean.

Safety signs:

A safety sign with a blue background and white writing signifies a mandatory – must do – instruction.

A safety sign with a yellow background and black writing signifies a warning – care and caution – instruction.

A safety sign with a red background and white writing signifies a fire equipment instruction.

A safety sign with a green background and white writing signifies safe conditions – for example, fire escapes, exits, first aid box and so on.

A safety sign with a diamond shape and either a red, blue, yellow, white or green background and black writing signifies that a package or load contains hazardous substances.

Please make a note of any points from this section that concern you.

Notes

HAZARDS AND RISKS

Take time to reflect on the question below, making some notes in the space provided.

List ten hazards that exist in your workplace, and rate their risk (chance of happening) as either low, medium or high. (Please give examples and be specific.)

1.

2.

3.

4.

5.

6.

7.

8.

9.

10.

PLEASE COMPLETE BEFORE CONTINUING

SELF-ASSESSMENT WORKSHEET

Please complete the following questionnaire, as honestly and accurately as you can. Rate your response to each statement or question on the following scale.

1 = Never; 2 = Sometimes; 3 = Usually; 4 = Often; 5 = Always.

1. I always consider safety issues before tackling a new job or task	1 2 3 4 5
2. I encourage co-workers to take safety seriously	1 2 3 4 5
3. I keep alert for unsafe conditions	1 2 3 4 5
4. I use safety guards and shields on machines and tools	1 2 3 4 5
5. I pace myself and avoid becoming too rushed or tired	1 2 3 4 5
6. I am careful when using electricity	1 2 3 4 5
7. I allow time to check things before I start something new	1 2 3 4 5
8. I ask for help or advice when I am not sure of something	1 2 3 4 5
9. I can accept constructive criticism about safety practices and my work	1 2 3 4 5
10. I input my ideas into the safety planning and programmes	1 2 3 4 5
11. I am particularly cautious when dealing with chemicals	1 2 3 4 5
12. I avoid taking short-cuts and cutting corners that may increase the risk of an accident or injury	1 2 3 4 5
13. I make sure that the area in which I am working is organized and tidy	1 2 3 4 5
14. If I feel tired or start to make silly mistakes, I take a break or change tasks	1 2 3 4 5
15. I avoid taking risks with other people's safety	1 2 3 4 5
16. I am particularly careful when lifting or moving objects	1 2 3 4 5

cont'd

17. I always try to take regular exercise and stay fit	1 2 3 4 5
18. I am careful with what I eat and avoid heavy meals during the day	1 2 3 4 5
19. I avoid using alcohol, drugs or excessive medication	1 2 3 4 5
20. I always use a seat belt when driving or am a passenger in a car	1 2 3 4 5

My score is _____/100 or _____%

Analysis

Between 80%–100%	Excellent.
Between 60%–80%	Very good.
Between 40%–60%	OK, but there is room for improvement.
Less than 40%	Watch out!

What safety points do you need to give some attention to?

Notes

____

____

____

____

____

PLEASE COMPLETE BEFORE CONTINUING

Self-assessment: Review

1. **I always consider safety issues before tackling a new job or task.** This is when you are at greatest risk, perhaps because your skill or competence is lower or untested. You may also be unaware of hazards and risks. Get trained, ask advice, think safety!

2. **I encourage co-workers to take safety seriously.** You can influence others, both by example and suggestion. Remember, it could be your safety which they are putting at risk.

3. **I keep alert for unsafe conditions.** Accidents usually leave clues and signposts, and nearly all can be prevented if these indicators are spotted earlier.

4. **I use safety guards and shields on machines and tools.** Your employer must provide them, and you **must** use them. It is not an option.

5. **I pace myself and avoid becoming too rushed or tired.** Research and our own personal experience tell us that we make more mistakes when we are tired or work hastily.

6. **I am careful when using electricity.** Even everyday appliances can kill. Follow instructions, use common sense and don't take chances.

7. **I allow time to check things before I start something new.** A few moments of preparation can save hours of regret.

8. **I ask for help or advice when I am not sure of something.** There is nothing wrong in admitting that you need help or don't know everything. The only weakness or failure is not asking for help when you need it.

9. **I can accept constructive criticism about safety practices and my work.** Safety is not just a personal issue – it can affect everybody. Keep an open mind and listen to the advice of others.

10. **I input my ideas into the safety planning and programmes.** You probably know as much about your job and the hazards and risks that exist, as anybody else. Input this knowledge and ideas when you can. If your manager does not regularly discuss safety issues, then make sure you encourage him or her to do so.

11. **I am particularly cautious when dealing with chemicals.** There are special regulations and controls for hazardous substances due to their high risk of causing serious injury or sickness. Make sure that you know and follow these regulations if you ever have to use chemicals in your workplace.

12. **I avoid taking short-cuts and cutting corners that may increase the risk of an accident or injury.** Short-cuts are usually short-cuts to disaster. Follow the rules, and keep to procedures.

13. **I make sure that the area in which I am working is organized and tidy.** It only takes one trailing cable, one blocked fire exit or one crowded storeroom to cause an accident.
14. **If I feel tired or start to make silly mistakes, I take a break or change tasks.** Accident investigators often find that fatigue and tiredness are contributing factors to many accidents at work.
15. **I avoid taking risks with other people's safety.** It is not just your health and safety which you gamble when you take risks, but also that of your workmates and colleagues.
16. **I am particularly careful when lifting or moving objects.** Over half of all injuries or accidents at work are caused by manual handling.
17. **I always try to take regular exercise and stay fit.** Prevention is better than cure. Look after yourself by eating well and keeping fit. This does not necessarily mean working out every day, just keeping active.
18. **I am careful with what I eat and avoid heavy meals during the day.** Large meals, particularly if eaten during the day, can lower our energy levels and slow down our reflexes.
19. **I avoid using alcohol, drugs or excessive medication.** Many organizations now prohibit the use of alcohol or prescribed drugs.
20. **I always use a seat belt when driving or am a passenger in a car.** The one time you don't may be the one time you should.

Make a note of any points in this review which you feel may be particularly relevant to your working practice.

Notes

Chapter 3 How to Improve Your Safety Rating

Now that you are aware of what hazards and risks exist, this chapter shows you how to improve your safety rating.

Before starting this chapter, please take a few moments to make a note of any ideas or actions in the learning diary and log in Chapter 1.

OVERVIEW

This section is designed to highlight some extra important areas and considerations when working on a day-to-day basis. Some of the points may be more relevant to some of us than others, but it is true to say that they must be considered by every employee, at one time or another.

In terms of both safety legislation and identifying hazards and risks in the workplace, certain areas have been identified as the most significant.

In a non-office workplace, or more industrial warehouse environment, there are several main areas that we need to consider. These are:

- manual handling and transporting
- slips, trips and falls
- display screen equipment
- hazardous substances
- fire
- noise
- plant and equipment maintenance
- electricity
- machines and safety equipment
- safe systems of work.

Together these comprise the most hazardous, or potentially most dangerous, areas in our workplace. Every year about 500 people are killed at work and several hundred thousand more are injured or suffer ill-health.

In addition to the penalties incurred by failing to meet legal, health and safety requirements, accidents in the workplace also have a cost. For employees, this cost can be in terms of money or lost wages due to having to take time off; for companies, accidents cost money in terms of materials, damage, insurance costs and, in some cases, lost customers or orders.

More attention to detail, and more attention to safety in each of the areas listed above, should reduce and probably eliminate most common accidents in the workplace.

Which do you consider to be the three most important aspects of health and safety applying to you?

Notes

SAFE MANUAL HANDLING

Have you ever experienced back pain or backache?

Has anyone you know ever had a serious back injury – perhaps a slipped disc or a trapped nerve?

Back and related pain is one of the most agonizing and acutely uncomfortable ailments which we can experience. Once damaged, the spine or back rarely fully recovers. Often the only short-term cure is rest – sometimes for weeks at a time. Remedial surgery is complex, expensive, risky and not always 100 per cent effective.

New legislation introduced in the last few years requires all employers to train all staff in the correct techniques and methods. It also requires employees to use the correct equipment, follow safety standards and advice, and, wherever possible, eliminate the need for lifting, carrying, stretching or twisting.

> *Back injury and backache are very common. It is important, first, that you know how to lift properly and, second, that you know what you can lift safely. Finally, it is absolutely imperative that, wherever possible, you use safety aids and equipment.*

In the UK, over 5.5 million working days are lost every year because of back-related injury.

Lifting and carrying

The most effective form of lifting and carrying is to do none at all – or as little as possible. Wherever possible, we must eliminate the need for carrying or loading. This means stacking items lower, using trolleys, breaking loads down into smaller boxes and so on.

In reality, we can safely lift much less than we sometimes think we can. The actual guidelines range between 15–25 kg for lifting safely from floor to table or waist level. This is assuming that the load is safe, secure and that we are lifting properly and holding the load close to our body.

The safest form of lifting is not to do any!

There is no safe weight if you are not lifting properly. When lifting, the most important points to remember are:

1. *Think before you lift – get smart! Prevention is better than cure.*
2. *Position yourself squarely in relation to the object and take a firm grip.*
3. *Keep your back straight and your head up.*
4. *Lift with your legs, not with your back.*
5. *Look where you are going all the time.*

Stretching and twisting

Stretching and twisting are both common hazards in the modern workplace. We seem to invent positions and work areas which make it difficult for us and do us no favours. Here are five things to consider to reduce or eliminate stretching and twisting.

1. Check your work area

Look at the location of your desk, chair, shelves, files and drawers. A small change can make big improvements.

2. Get a good posture

Avoid slouching and make sure your chair is adjusted properly.

3. Check your seating

Is the chair sturdy? Does it support your back? Can it swivel and turn? Is it on castors? Make sure that you do not tilt or rock – losing your balance can lead to a nasty fall.

4. Do not overload shelves and desks

Apart from making it difficult and time-consuming to find things, overloaded shelves and desks make an accident more likely and can turn an innocent storage rack or load pallet into a potential hazard.

5. Use stepladders and be sure-footed

Although such actions are clearly not common-sense, if you look around most workplaces you will see people standing on chairs – even swivel chairs with castors – desks, cardboard boxes and all manner of objects to reach high places. Always stop, think and use the right equipment.

Repetitive strain injuries

Other common hazards are the manual movement of loads and frequent forced or awkward movements of the body. These can often lead to back injuries and severe pain in the hand, wrist, arm or neck, sometimes known as repetitive strain injury (RSI).

A movement or activity can still be potentially hazardous even though it might involve only very light lifting or slight twisting. What makes it potentially hazardous is the repetitive nature of the task. For example:

- As till price scanners have increased in speed, supermarkets have had to carefully assess the checkout operators' position and work routine.
- Telephone operators, especially those who need to use a keyboard or write while talking on the telephone, may require headsets to avoid injury caused by balancing the handset between their neck and shoulders in order to leave their hands free.
- Keyboard operators with a heavy data input role or who need to make excessive use of a mouse may need to vary their task or take regular breaks to avoid injury to the wrist and hands.
- Packing and production line staff who perform the same task hundreds of times per hour need to vary their task to avoid putting continuous, excessive strain on one area of the body.

Thousands of 'hidden' injuries are caused by the repetitive and trivial tasks that we all do every day. Aches, pains, bruises, stress, tiredness and soreness are all largely unreported.

Safe mechanical moving

Moving materials mechanically is also hazardous, the principal risk being that of material falling from a height or a moving device and either crushing or striking people. Every year there are over 5000 reported accidents involving transport in the workplace, and there may be many times more than this number that are never reported. Of these accidents, over 60 result in a death. Safe handling and transporting can be summarized by three key points.

1. Safe lifting

You must avoid lifting in the workplace wherever there is a risk of injury. You must also fully assess the risk of injury and take all practical steps to reduce that risk when lifting, shifting or moving objects.

If possible, always consider automation or mechanization as an alternative. You must also know how to lift safely, and this is a very important obligation. If you do not know, or are not sure how to lift certain objects or certain sizes and shapes of object, it is your responsibility to ask and find out.

2. Safe stacking

Materials and objects must be stored or stacked so that they are unlikely to fall and cause injury. Always stack on a firm level base. Use a properly constructed rack when needed and secure it to a wall or floor if possible. Use containers correctly and make sure that pallets or racks are suitable for the job and not damaged. Always ensure that stacks are stable, and never exceed safe loads that are specified, or are part of your lifting regulations. Check that items do not protrude from stacks, and do not climb racks to reach upper shelves; always use ladders or steps.

3. Safe transporting

When transporting any goods you must lay out your workplace so that pedestrians are safe from vehicles, and properly train all drivers. Where practical, always separate vehicle operation areas and pedestrian walkways. Where they must cross, you must provide clearly marked pedestrian crossings.

You must control pedestrian access to loading bays and delivery points, and ensure that drivers can see clearly and that pedestrians can be seen and are aware of vehicles. Where necessary, consider the use of mirrors, high visibility clothing, audible alarms and lighting in both the workplace and on vehicles. Once you have proper rules in place it is extremely important to make sure that all drivers follow these rules, including visiting drivers. Vehicles should be checked daily and have faults rectified promptly. Vehicle movements should be supervised, particularly when reversing near blind corners. Drivers should always use recognized hand and arm signals.

Take a moment to make some notes regarding steps to improve safe manual handling and mechanical moving in your workplace.

Notes

ACCIDENT PREVENTION

Slips, trips and falls

As you have seen from the statistics, these are a very common form of accident and injury.

In an average year there are nearly 35 000 accidents caused by slips, trips or falls at ground level. These types of accident are most likely in the following four areas.

1. **Stairways.** Stairways should have handrails, access guards and safety treads and must be well lit.
2. **Carrying.** Do not try to carry too much and, in particular, do not carry a load that you cannot see over or around clearly. It is far better to make two trips than one trip and have an accident.
3. **Rushing.** Do not rush or hurry, especially around corners. As well as running the risk of bumping into somebody else, you also increase the chance of yourself tripping or slipping, and may not see hazards around you.
4. **Think ahead; look ahead.** Look for uneven flooring, steps, kerbstones, cables, chairs, corners of desks and sharp objects. Remember, there is no such thing as an accident. Accidents are preventable occurrences caused by neglect or lack of forethought.

Keep floors clean, level, unbroken and non-slip. Keep gangways clear and unobstructed, and exits properly marked.

Cuts and grazes

Associated with slips, trips and falls are cuts and grazes. Here are some points to note.

1. **Pointed objects.** Keep pointed objects separately in a drawer; do not leave them on desks or in pen pots.
2. **Staples and staplers.** Use staples and staplers carefully, particularly when removing staples from envelopes or boxes.
3. **Paper and boxes.** Be careful when handling paper and boxes. A paper cut can be very nasty and quickly turn infectious. Be alert for staples and other sharp objects inside boxes that you may not see.
4. **Drawing pins.** Do not leave drawing pins, blades or paper clips lying around. Whilst innocent enough, all of these are potentially dangerous objects and can cause injury, cuts and grazes.

5. **Broken glass.** Most importantly, do not put broken glass or sharp objects in wastebins. These must be disposed of separately and, if possible, kept well away from ordinary waste. If you do need to dispose of these in a dustbin, be sure to put them in a secure cardboard container and mark accordingly.

Working from heights

> *Killed by a ten foot fall: a stage fitter fell just ten feet, breaking his back and never regaining consciousness. The accident investigation highlighted that a safety harness should have been provided; this would have prevented his death.*

Working from heights is a particularly hazardous activity. Always make sure that you:

1. **have been properly trained**
2. **follow procedures**
3. **wear safety equipment and harness**
4. **never take chances.**

Ladders and scaffolding must be suitable for the purpose and correctly and securely positioned. The manufacturer's instructions must be followed with regard to loading and erection.

Take a moment to make a few notes on how you could improve your workplace to avoid slips, trips and falls, cuts and grazes and falling from heights.

Notes

____

____

____

____

____

DISPLAY SCREEN EQUIPMENT

The introduction of computer technology has been so rapid and so widespread that long-term health effects are virtually unknown. We are the 'guinea-pigs'.

The use of computers over the last 10–15 years has multiplied. This has created a new need for safety awareness, and recently introduced legislation requires new and higher standards of both manufacturers and operators in the use of this equipment.

Are you a habitual user? The commonly accepted guidelines suggest about two to three hours' cumulative usage per day. However, certain types of work might be more intense and be classified differently. If you use the computer to this extent then you should have your work station formally assessed once a year. Even if you use a computer less than this, it makes sense to apply safe working practice.

Nowadays, many of us spend some time in front of a computer screen. This will probably increase. This particular safety hazard did not exist to any degree 10 to 15 years ago, so recent legislation has been introduced to safeguard and check the safety of display screen equipment. Here are five guidelines for making sure that your computer screen or VDU terminal is safe to use.

Guidelines to working safely

1. Check that your work station is correctly configured

Your manager, supervisor or safety representative can advise you on how best to set up and configure your work station or terminal not only to comply with safety regulations but also to make it as easy and as stress-free to use as possible.

2. Check your posture

Incorrect posture greatly increases the risk of injury, accident, sickness and aches and pains. As we may spend several hours of our working day in front of a computer screen, we must make sure that we sit properly to minimize the occurrence of aches, pains or physical strain.

Because of the relative sedentary nature of the work of computer operators, posture mistakes can quickly lead to cumulative pain and distress.

3. Make sure that your work station is clear and tidy

Clear a space around the computer so that you can adjust the keyboard, see the screen, and not be distracted or affected by glare. These can all improve the ease of use, and ultimately improve the safety, of your computer terminal or display screen.

4. Vary your work routine

Build regular breaks into your work routine. Many of the injuries and illnesses due to using computers develop over many years. Repetitive strain injury, known as RSI, is a term describing injury to tendons, muscles and joints caused by repeated actions over a long period of time.

Whilst this condition is very new, and research is still continuing, experts believe that the risk of RSI can be greatly reduced if we take a break every hour or so for just a few minutes. This may not mean going off and having a cup of coffee, just simply changing activity. Do some filing, look up some information, go to see somebody, make some telephone calls – anything that takes you away from the computer and is a change of activity.

5. Minimize glare and have adequate lighting

One of the biggest hazards of computer usage is glare from the screen. This may be due to an incorrectly adjusted VDU or could be due to reflected light from the screen, caused by badly placed fluorescent or other lighting, or glare from windows and sunlight. Adjust your screen to minimize glare, and use blinds, if necessary, at certain times of the day. Also make sure that you have adequate lighting, both on the work that you are reading and in your general work area.

Despite all the attention paid to glare and radiation from monitors, there are potentially more dangers from the output of ozone from laser printers, noise from line printers and reflected natural or fluorescent light.

6. Minimize noise

Noise, whilst not usually a problem in most modern offices, may be a consideration in factory, production, packing or other environments, making it difficult to concentrate and causing headaches or tiredness. The use of screens or partitioning may improve matters. Also, the use of line, daisy-wheel or similar impact-based printers may require acoustic screening or movement away from the main working area.

Make a few notes on how you could improve your VDU or computer work station.

Notes

CHEMICALS AND HAZARDOUS SUBSTANCES

All chemicals or substances that may cause a threat to health must be clearly labelled, stored and controlled. You should have a good working knowledge of any such chemicals that you may use or come into contact with.

This working knowledge should include:

- the nature of the substance
- the nature of the associated hazards and risks
- how to recognize symptoms of chemically induced injury and illness
- what to do in the event of accident or injury
- protective equipment that should be worn
- how to store and maintain the chemicals
- the legal requirements controlling the substances

Most importantly, all chemicals and hazardous substances should be regularly checked and inspected.

Read the labels! There are dozens of hazardous substances in our homes and workplaces. These range from cleaning materials to headache tablets, petrol to air fresheners. Always follow instructions and know what you're working with. If in doubt, don't use it.

Make a list of the hazardous substances that exist in your workplace or operations and ways that you could further improve safety standards.

Notes

FIRE

An outbreak of fire, although fortunately a rare occurrence for most people, is usually devastating when it takes place.

Good fire safety practice applies to the general work environment and work space. Whilst most of it may be common sense, it is easy to overlook some aspects.

Whilst the risk of fire breaking out may be low, it is a major hazard and one that can produce terrible consequences. Workplaces tend to be busy with people, and therefore fire safety is a very important consideration.

Fire is one of the most destructive and devastating things that it is possible to imagine.

Besides the danger – the very real danger – of loss of life, a fire can put a workplace, factory or warehouse out of action for months, years and sometimes permanently. Fire has been known to bankrupt businesses because, even though insurance will cover the cost of the damage, it will not cover the cost of lost customers, business or production.

Keys to fire safety in the workplace

1. Know your fire procedures

You have a legal obligation to read all fire notices and keep up-to-date with fire procedures. You must brief new staff, and if you yourself have not been briefed then you must ask your manager or supervisor. Although fire drills may take place when it is least convenient or when it is raining, you must take them seriously. Always treat a fire drill as though it is the real thing – one day, it might be.

2. Know what to do in the event of a fire

Do you know what you would do if you discovered a fire in your workplace? What would you do first?

- Evacuate the building?
- Sound the alarm?
- Dial 999?
- Try and put it out yourself?

Although all the above are good ideas, the order and sequence of your actions are important. First you must sound the fire alarm, which will ensure that the building or area is evacuated safely. If it is a very small fire in a very small contained area – for instance, some packing material or a wastebin on fire – you have been trained to use the appropriate fire-fighting equipment and it is within your own fire safety procedures, then you may wish to tackle the fire on your own. However, in most cases, the safest thing you can do, after sounding the fire alarm, is to dial 999, and evacuate the area as quickly and safely as possible, being sure to close all doors and windows.

3. Fire prevention

This is clearly the area on which to concentrate the most effort. You should understand not only how a fire starts but also the most significant hazards and risks existing within your workplace that have the potential to cause fire. If your workplace allows smoking, then this will be the greatest fire hazard. Never empty ashtrays into wastebins, and always keep cigarettes away from paper and combustible materials. If your workplace is a non-smoking area, then the greatest fire hazard may be electrical appliances or heaters. Again, these must always be kept away from paper and other combustible material.

Fire checklist

- ☐ **Know your fire extinguishers and their locations.**
- ☐ **Know the locations of fire exits and fire doors.**
- ☐ **Never wedge open fire doors.**
- ☐ **Fit and check smoke detectors.**
- ☐ **Be careful when siting radiators or heating fires.**
- ☐ **Be especially careful with cigarettes.**
- ☐ **Store flammable liquids and gas containers correctly.**
- ☐ **Take fire drills seriously.**
- ☐ **Know where the fire alarms are and how to use them.**
- ☐ **In the event of an alarm or fire breaking out, leave the area quickly and calmly.**
- ☐ **Empty bins and dispose of rubbish promptly.**

Make a few notes on how you could improve the standards of fire awareness and precautions in your work place.

Notes

NOISE

Noise must be carefully monitored and controlled in all work environments, but especially in those areas where loud noise is more common. Proper risk assessments should be carried out at regular intervals to measure the noise levels, and to assess any associated risks.

Safe noise levels

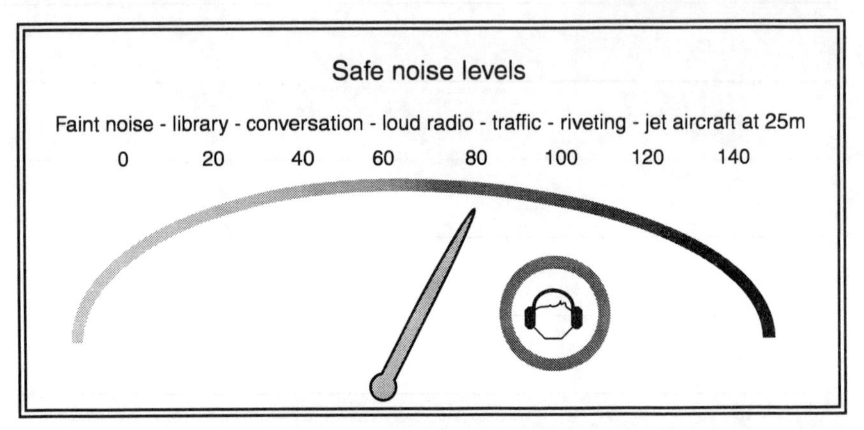

The chart above shows the decibel range of noise from zero, which is just a very faintly audible sound, through to 140 decibels, which you would experience when standing 25 metres away from a jet aircraft taking off. At this upper limit you would experience extreme pain.

An example of nuisance noise levels would be trying to make a telephone call from a motorway breakdown telephone. You will find that you will have to shout at the top of your voice just to make the other person hear what you are saying, and hearing the other person on the telephone would be extremely difficult because of the traffic noise.

If you were to be exposed to this level of noise continuously or regularly during your working day, it would not take very long for your hearing to suffer. It is also important to remember that loss of hearing may be very gradual. Unlike some other health and safety accidents and injuries, it may take years before you begin to notice the effects, by which time it is too late to do anything about it.

How to protect your ears

First, you must investigate all practical ways in which the noise levels can be reduced or loud noises eliminated altogether. This can be achieved through changing equipment, using quieter machines, or possibly even eliminating the whole process altogether. It may also mean working with suppliers to find new ways of reducing noise from equipment. If buying new machinery, noise levels should be an important consideration.

Ear muffs

Protective ear equipment must be worn in areas where noise levels may exceed 90 decibels. This would be louder than the engine of a heavy lorry at close quarters or just slightly noisier than a very busy street.

Noise enclosures

Silencers and other protective aids can also help to cut down the noise from machinery and equipment. The manufacturer or supplier of all equipment is obliged to provide information regarding noise levels, although this is no substitute for testing the noise level yourself.

Is noise a problem in your workplace? Make some notes on actions you can take or points to check to further reduce noise or protect the hearing of you and your colleagues.

Notes

___

___

___

___

___

___

___

___

___

PLANT AND EQUIPMENT MAINTENANCE

When maintaining or repairing any plant or equipment, special care and consideration is required. Maintenance is carried out to prevent problems arising and to put faults right. We must make sure that we do not create more problems in trying to do so.

Extra care is needed when climbing and working at heights, or when doing work that requires access to unusual parts of the building. Hazards can arise when:

- working on machinery – including accidental or premature start-up
- using hand tools or electrical equipment
- during contact with materials that are normally enclosed
- entering vessels or confined spaces where there may be toxic materials or a lack of air.

There may also be temptations to cut corners and rush through tasks in order to get equipment up and running again as quickly as possible. However desirable this may be, it is vital that health and safety considerations are not neglected in the process.

Steps for safe maintenance working

1. Create a safe working area

It is important that you always create a safe working area before you start. Ask yourself:

- Is the job really necessary?
- Can it be done less often without increasing other risks?
- Should it be carried out by specialists?

Never undertake work for which you are unprepared or not competent. Always plan the work to cut down the risks – for example, the difficulties in coordinating maintenance with routine work can be avoided if maintenance is carried out before start-up or after shut-down. Also, remember that access is safer if equipment is designed with maintenance in mind.

2. Make plant or machinery safe

Always make plant or machinery safe before starting work; never risk working on live machinery. Most maintenance should be carried out with the power off, and you may need to isolate electrical and other power supplies. If the work is near overhead electrical conductors – for example, close to overhead travelling cranes – the power should be cut off first. Also, isolate plant and pipelines containing pressurized fluid, gas, steam or other hazardous material. Isolating valves should also be locked off.

Some parts of the plant may need to be supported, to avoid the risk of falling. You should check that all mobile plant is fully stationary – that is, in neutral gear, with the brakes applied and the wheels blocked – and that components which operate at high temperatures have had time to cool. Clean and check all vessels or equipment that use or contain toxic or flammable materials before the work starts.

3. Hand tools

You must ensure that hand tools are properly maintained. For example, check that hammers do not have broken or loose shafts and worn or chipped heads and that heads are properly secured to the shafts. Files should have a proper handle; never use them as levers. The cutting edge of chisels should be sharpened to the correct angle. All split handles are dangerous.

Use tools only for the purpose for which they are intended. Screwdrivers should never be used as chisels and hammers should never be used on them. Avoid using tools that are not designed for the job in hand, and make sure that all tools are returned promptly to the tool box or work storage area.

4. Working in confined spaces

This is a high-risk activity. In the UK about 15 people every year are killed while working in confined spaces. Many more are seriously injured. Asphyxiation and toxic fumes are the two most common causes of death, but others include drowning in free-flowing solids – such as grain in silos – and death caused by fire and explosions. Unless you have been trained properly, and are fully equipped to enter the confined space, do not do so. Wherever possible, eliminate the need to work in a confined space by investigating whether the work can be done from outside.

Precautions for working in confined spaces include:

1. *isolating dust and fumes*
2. *testing the atmosphere*
3. *using non-sparking tools*
4. *having rescue equipment and trained staff at hand*
5. *safe and adequate lighting.*

5. *Vehicle repairs*

Vehicle repairs are also a particularly hazardous activity, and adequate precautions must be taken to reduce all possible risks. When repairing the vehicle always make sure that the brakes are applied and wheels are obstructed. Always start and run engines with the brakes on and in neutral gear. Vehicles should be supported on both jacks and axle stands; never rely on jacks alone. Raised bodies, such as lorry cabs or trailers, must always be propped when working. Be constantly aware of the risk of explosion when working on vehicles, especially when draining and repairing fuel tanks. Battery gases and acid also pose a particular hazard. Be sure to use protective equipment when it is required.

Garage and motor workshops are full of hazards: fumes, chemicals, electricity, hand tools, lifting, working in confined spaces, heavy lifting equipment, moving vehicles.

Be especially safety-conscious – always.

Make a note of any points from this section that concern you.

Notes

ELECTRICITY

The trouble with electricity is that you can't see it, you can't hear it, you can't smell it and, when you feel it, it's usually too late.

All electrical work and installations must be carried out by suitably qualified and trained staff. Yet, even when this rule is followed, over 1000 electric shocks and burns are reported per year, of which about 30 are fatal. However, there are six very simple keys to working safely with electricity.

How to minimize the risk of electric shock

1. Isolate

Unless absolutely necessary, all equipment must be isolated before starting work on it. Electricity cannot hurt you if it is not there.

2. Reduce

Reduce the voltage wherever possible, and consider using hand or air-driven tools. Portable tools can run at 110 volts from an isolating transformer.

3. Use circuit-breakers

Use safety devices, both at the main power source and for each hand tool appliance or machine.

4. Make installations safe

Create and maintain safe installations. This means never overloading sockets or extension leads and using the correct fuses, circuit-breakers and earth devices. There must be a clearly marked switch or isolator near each fixed machine to cut off power in an emergency. All mains switches must also be readily accessible and clearly identifiable.

5. Beware of overhead power lines

Contact with overhead electrical lines accounts for over half of the fatal electrical accidents each year. Electricity can flash over from overhead power lines even though plant and equipment may not touch them. Never work where any equipment, such as ladders, a crane jib, tipper lorry or a scaffold pole, could come within nine metres of a power line, without seeking advice. Always consult your electricity company first.

6. Maintenance

All electrical equipment, wiring installations, generators or battery sets – and everything connected to them – must be maintained to prevent danger. This means carrying out regular checks and inspections, repairing and testing as necessary. How often you do this will depend on the equipment that you use and the way in which you use it.

Using electrical equipment and appliances

Whilst using electrical equipment is largely a matter of common sense, and should be fairly familiar since we all use electricity, there are five points that should always be considered if we are to treat electricity with due respect.

1. Do not overload power points

When new equipment, computers and other appliances are introduced into older workplaces, there can quite often be a shortage of power supplies or power points. However, this is no excuse for unsafe working practice. Always insist on having new power points installed by a competent person. Adapters, whether they are the plug-in variety or trailing multi-sockets, are potential hazards and can increase the likelihood of accidents through fire, falling or electrocution.

2. Use the correct fuses

An ordinary plug can take a variety of different fuses. These range from 3, 5 and 13 amp, and there are other sizes as well. The type of fuse that you will need will depend on the appliance which you are using. Be sure to check this or, better still, have someone who is competent do this for you.

3. Replace worn or damaged wiring

This is a very obvious point, but it is not uncommon to find frayed, damaged, snagged, torn or even scarred wiring in a workplace. Do not take the risk.

4. Take precautions with heaters and fires

The most dangerous form of heater is the radiator type of fire in which a hot element or hot air is radiated or is blown from a portable appliance. These heaters must be kept well away from any combustible materials, and switched off and isolated periodically during the day – especially at the end of the working day.

5. Correctly install and repair

Always make sure that all electrical equipment is maintained, installed and periodically inspected by a competent person. Every electrical appliance should be inspected on an annual basis, and a written record kept of this. Under no circumstances attempt DIY repairs of any electrical appliances. Apart from possibly breaking the law, you could be dramatically increasing the chances of an accident – one for which you may be responsible.

Make a note of any points from this section that concern you.

Notes

MACHINES AND EQUIPMENT

The modern workplace is full of technology and machinery. Consider how little of your job you could do if there was a power cut. You could not operate machines and equipment or send faxes, most likely the telephone system would not work and the computers, photocopiers and the heating would probably go off. This reliance on electrical power means that the workplace is full of potentially dangerous equipment. However, none of this equipment is dangerous unless we use it incorrectly.

Always check for the following:

- *loose electrical connections – whether these are plugs, wiring or connections*
- *exposed leads and cables – not only those trailing across the floor where they risk being pulled out and potentially causing an accident, but also those with worn or twisted wiring.*

Many serious accidents at work involve machinery. Amongst other hazards, hair or clothing can become entangled in rotating or moving parts. It does not take much imagination to realize that the many, regularly occurring, horrific accidents and injuries are caused by the misuse of machines and equipment at work.

Minimize the risks

The first stage is to assess all risks very carefully. Think about all the work that has to be done on a machine and what risks may be involved in this. Consider the experience or training of the people using the equipment or machines. What opportunities are there for people to behave foolishly or carelessly or make mistakes? Always consider all the hazards that may exist within a piece of machinery.

Safety guards should be made available wherever a hazard or risk exists. Dangerous parts should be enclosed within fixed guards and, if practicable, must be fixed firmly in place. Make sure that you are using the highest-quality materials. Do they allow ease of use, clear vision and yet still protect operators from injury? You may need to have trip systems, pressure-sensitive mats or automatic guards built into electronic or automatic machines. You may also need to consider working with your suppliers to improve the safety and safety guards on your machinery.

Failure to use safety guards is the most stupid, irresponsible and dangerous thing you can do . . . and yet every day people are maimed, blinded and injured as a result.

Once you have assessed the risks and made sure that adequate safety guards are in place, it is then of paramount importance that the machine operates safely. Even the most innocent-looking machines can cause very serious accidents and injuries. People using equipment with moving blades, such as guillotines, cutters or slicers, must receive appropriate and adequate training. Some machines that are prescribed as particularly dangerous can only be used in factories by young people – that is, people under 18 but over 16 – after full instruction and sufficient training and then only under close supervision.

Proper protective equipment must be provided and worn if required. Also check that adequate lighting is in place to allow safe operation of this machinery.

All machinery should be regularly checked and maintained by a properly trained and qualified person. Maintenance work should not be carried out while machinery is still in operation, as this can present additional hazards.

When any of the equipment or machinery in the workplace – for example, the photocopier – stops working, as it occasionally does, resist the temptation to try to fix it yourself. You are dealing with very sophisticated and potentially lethal equipment. Although it might seem a good idea to poke round the photocopier with a ruler or take the back off the computer because you think you know what the problem is, you may actually be causing more problems than you solve. Call in an expert every time.

The final factor that can turn a normally safe piece of equipment into a potential hazard is faulty operation. So the next time the photocopier starts to make a funny noise, or you notice that a piece of equipment or hand tool is either hotter than normal or smells of burning, consider these messages as telling you that the risk of accident and injury is increasing.

The solution? Simply turn off the piece of equipment, disconnect it from the mains and call someone who can do something about it. Do not risk the consequences.

Devise safety checklists

It is important to create checklists for all operators to use. Here is an example of what a simple checklist may cover:

1. **Do you know how to stop a machine before you start it?**
2. **Are all the guards in position and all protective devices working?**
3. **Is the area around the machine clean, tidy and free from obstruction?**
4. **Have you informed your supervisor if you think that your machine is not working properly or any safeguards are faulty.**
5. **Are you wearing appropriate protective clothing and equipment, such as safety glasses or shoes?**

The checklist may continue with a series of appropriate 'do not' instructions – for instance:

- **Do not wear loose clothes.**
- **Do not wear rings.**
- **Do not wear dangling jewellery or chains.**
- **Do not wear your hair long and loose.**
- **Do not distract people who are operating machinery.**

Make a note of the most important points from this section.

Notes

SAFE SYSTEMS OF WORK

All the previous sections have looked at particular hazards. Now we need to think more generally about safe procedures and systems at work and how to carry them out.

How to create safe systems of work

1. Have clear procedures

It is important to have clear procedures. People do not plan to fail; they fail to plan.

Procedures help you work correctly and safely. For activities which carry serious hazards and risks it is worth writing them down in the form of, say, a written 'permit to work'. This may not be necessary in many ordinary jobs, however. When looking at your systems do not forget routine work as well as less routine work, such as maintenance.

Also consider procedures to deal with emergencies such as fire, spillage or plant breakdown, taking into account all the different tasks that people do. Ask those who do each job to tell you what they do and how they do it, so that they can help you identify the hazards and risks.

2. Have safe procedures

Having clearly communicated procedures is not enough; they have to be recognized as safe procedures. Therefore somebody with responsibility, as well as the knowledge required, must assess the procedures to see if they are achieving their aim.

In the final analysis, good health and safety is about two things: having safe systems of work, and following them to the letter.

---

3. Cultivate the right attitude

Attitude also plays a significant part in any safety system. You cannot rely solely on your systems as always being right. Check that your rules and procedures not only deal with all the risks, but are also being followed, particularly if people are working outside normal hours with less supervision that usual. Check that people are following procedures, or whether they only do this when they are being checked. Are there any training gaps, and is the level of supervision right for the staff involved?

> ***Regular review:***
>
> *Once you have systems in place, it is important to check and improve them at regular intervals, especially if the tasks or the operations change significantly. These changes would, of course, alter the hazards and risks associated with them.*

4. Learn from mistakes

Accident reporting and investigation must also be written up in a series of emergency procedures, and these should be in place before accidents and emergencies occur – not after. If your job involves supervision of others, or you have special responsibility for particularly hazardous work areas or operations, you must be fully aware of these procedures and know who needs to be involved in the event of any accidents or emergencies.

As well as being a legal obligation, it is absolutely vital that the knowledge gained from any accidents, injuries or other unplanned events is learned from and applied to prevent any reoccurrences.

Every accident, incident or near-miss must be carefully evaluated, not just to apportion responsibility but also to learn and improve.

Make a note of any points from this section that concern you.

Notes

Safety Checklist

Make a list of **everything** that you might consider when assessing your work environment. You might find it useful to divide your list into each of these four categories.

1. Plant and equipment	**2. Tasks and operations**
3. The work environment	**4. The individual**

PLEASE COMPLETE BEFORE CONTINUING

Chapter 4
Attitude Makes the Difference

This chapter shows how your attitude can either contribute to safety or cause accident and injury in the workplace.

Before starting this chapter, please take a few moments to make a note of any ideas or actions in the learning diary and log in Chapter 1.

THE COST OF POOR SAFETY

We all have plenty of reasons for learning to be safety-conscious and we all know the consequences of poor safety and some benefits of good safety. Of the two, which would you think is the greater motivation – the consequences of poor safety or the benefits of good safety?

In fact, the consequences of poor safety tend to motivate us slightly more than the benefits of good safety. For instance, imagine that you are chopping an onion at home. The thought of slicing into your finger would probably motivate you more to be careful than the benefits of having a neatly sliced onion. So let us consider the cost of poor safety.

What does safety really cost?

1. Discomfort and pain

When accidents happen to us or we injure ourselves either at work, in the home or anywhere else, we generally experience discomfort and pain. Quite often, this pain is extreme. Anyone who has ever slipped a disc, dislocated a shoulder, broken a leg or suffered serious cuts and bruises will appreciate the degree of pain involved. We tend to learn very quickly after any one of these experiences.

2. Work disruptions

Accidents in the workplace can cause all kinds of disruption to our normal work schedule. An accident during the day might mean that we waste hours of working time while we receive medication. The fact that we cannot carry out our normal duties may then have a snowball effect, stopping other people working and disrupting all kinds of activity – orders may be delayed, typing may not be done, schedules may be missed and appointments postponed.

With a more serious injury, we may have to take time off work. We may not be able to play our favourite sport or do our favourite hobby. We may have to postpone holiday plans. All these are consequences of not taking health and safety seriously enough. In many situations – for example in offices – instead of accidents, people experience stress, bodily aches and continual tiredness; this means that they are less effective both during the working day and at home.

3. Money

Most organizations are very concerned with money. Accidents, injuries and even near-misses cost money. This cost can be in terms of actual damage to machinery, lost wages, higher insurance premiums, lost customers and possibly even compensation payments made to employees or customers.

4. Lost productivity and business

Accidents and injuries can lead to lower productivity, an increased workload for others, missed orders and inconvenienced customers – none of which is good for business.

So you can see that the consequences of poor safety are quite considerable. We have all read about the major disasters that have happened in the last few years. Whilst train crashes, fires in underground stations, sea and river disasters all grab the headlines, every day there are thousands of accidents in workplaces just like yours that cost millions of pounds and many hours of discomfort and pain.

Make a note of any costs of poor safety that you have experienced or witnessed in your organization.

Notes

YOUR ATTITUDE MAKES THE DIFFERENCE!

It seems that accidents do not just happen. Something causes them, and that something is usually people. To be more specific, it is often people's attitudes that cause accidents, but how do we define attitude or, more importantly, a safe attitude?

There are four keys to having a positive safety attitude.

1. Be informed

Take time to understand and think about the hazards and risks that exist, not only in your own work environment, but also in the various tasks and jobs you undertake during the day.

2. Be aware of safety procedures

Every organization has safety rules and procedures, including yours. It is very important that you learn the rules concerning health and safety and then keep to them. Not only is it clearly unsafe to break safety policies and procedures, you may actually be breaking the law and therefore be liable to prosecution.

Attitude Makes the Difference

- ***Be informed.***
- ***Play by the rules.***
- ***Cooperate.***
- ***Be alert, awake and aware.***
- ***Think safety.***

3. Cooperate with safety representatives

The Health and Safety at Work Act states that it is the responsibility of every employee to cooperate fully with safety representatives and managers who are implementing safety policy and procedures. To put it more positively, you will benefit the most from good safety procedures, so it makes sense to cooperate with other people around you who are working towards that goal.

4. Be alert

Accidents happen when we walk around with our blinkers on, so always be alert, awake and aware of what is going on around you and of what hazards may exist, or potentially exist, in all situations.

Attitudes that cause accidents and injury

Now that we have considered some positive attitudes that can help us prevent accidents and injuries, let us look at the more negative attitudes that can cause accidents in the first place.

Attitudes that Cause Accidents

- ***Overconfidence***
- ***Laziness***
- ***Stubbornness***
- ***Impatience***
- ***Ignorance***
- ***Showing off***
- ***Forgetfulness***

1. Overconfidence

Overconfidence means thinking that accidents cannot happen to us, that they only happen to other people and believing ourselves to be 'too clever' or 'too good' at what we do.

2. Laziness

There is a saying that states 'A short-cut is only a fast route to a shortcoming'. Do not try to cut corners or take unnecessary risks. You may be risking your own life as well as somebody else's.

3. Stubbornness

Many jobs require people to wear personal protective equipment (PPE), such as hard hats, ear protection, special clothing, safety boots and so on. In fact, we all wear a particular form of protective equipment each time we make a journey by car – our seat belts.

However, consider how many people wore a seat belt for every journey before it was made law? Statistics show that it was very few of us, and the reason is probably stubbornness and bad habits. Even though we all understand it is better to wear a seat belt than to fly through the windscreen in the event of an accident, it took an Act of Parliament to make people do so for every journey.

4. Impatience

Many accidents are caused by people trying to do a task too quickly without paying due care and attention to what is going on around them or the consequences of not carrying it out properly. Very often we end up having to do something twice because we did not do it properly the first time. As the old saying goes: 'More haste, less speed'.

5. Ignorance

Clearly, if someone is ignorant of the dangers inherent in any particular operation or task, it might be unfair of us to expect them to know how to behave safely. However, the law says that ignorance is no defence.

When a Health and Safety Inspector investigates a company after an accident or injury, the company must prove that its employees were sufficiently trained and had the right knowledge and skills for the jobs and operations that they were carrying out.

Ignorance is no defence when it comes to health and safety, so make sure that you really understand and know the hazards and risks that exist in your workplace.

6. Showing off

A moment's levity can lead to long-term regret. We have all occasionally done something foolish and, fortunately, probably escaped too serious a consequence. However, always be aware that when we let our guard down we increase the risk of accidents.

7. Forgetfulness

We are all probably guilty of this from time to time. Even when we know what we should be doing, even with our normal best intentions, occasionally we forget. Safety is a full-time job, and it requires your full-time attention. Always consider safety; never forget its importance.

Make a note of any bad attitudes to safety which you have noticed either in yourself or your colleagues.

Notes

CAUSES OF ACCIDENTS AND INJURY

Let us look at the six key 'attitude' causes of accidents and injuries in offices.

1. Carelessness and bad habits

There is nothing particularly sophisticated about this, but it is the cause of more accidents than almost all the other factors put together. Avoid carrying out tasks wrongly or carelessly. There's a saying in Health and Safety which states that 'a casual attitude produces a casualty'. Just make sure that you are not that casualty.

2. Breaking the rules

When you know what to do, make sure that you do it.

3. Not knowing the rules

The law states that it is the organization's responsibility to train everybody – including suppliers, contractors and customers that may come into contact with you – in safety rules, policies and procedures. If you do not know, it is **your** responsibility to ask.

4. Ignoring faulty equipment

Whether it is a faulty chair which gives us backache, or a faulty computer which gives us an electric shock, anything mechanical, moving or electrical has enormous potential to cause us harm and needs to be treated with great respect. If any equipment which you use becomes faulty in any way, then it is your responsibility to stop using it, isolate it and tell someone about it.

5. Not thinking safety all the time

Accidents do not take coffee breaks nor do they take days off. The next accident could be a year or three seconds away – we never know. We must never – any of us – stop thinking safety all the time.

6. Thinking it cannot happen to us

Whenever you think this, you are at your most susceptible to accident, injury or ill-health. Of course, the truth is that an accident can happen to anybody at any time, to any one of us, any day. We must constantly be aware of the hazards and risks in our offices and how we can minimize those risks and make it safer and healthier for everybody.

Make a note of what you have learnt from the last two sections.

Notes

BE SAFE!

- ☑ **You have most to gain from working safely.**
- ☑ **You have most to lose.**
- ☑ **You are responsible for your own safety.**
- ☑ **All accidents can be avoided.**
- ☑ **All hazards should be reduced or moved.**
- ☑ **Risks must be minimized at all times.**
- ☑ ***Don't take chances.***

Which of these attitudes do you recognize in yourself or others?

Notes

20 Ideas for a Safer Workplace

Mark the five ideas that you want to improve on or implement.

1. Always consider safety issues when working or walking around the workplace.
2. If you see someone who is doing something that is **not** particularly safety-aware or is dangerous, point it out to them.
3. Keep alert for things that are broken or poorly maintained and get them fixed quickly. These are accidents waiting to happen.
4. Be careful not to overload plug sockets with adapters and extension leads. This is breaking safety rules.
5. Try to balance different types of work, or at least have a regular break or change of scenery at breaks.
6. Keep work areas clear and uncluttered.
7. Accept constructive criticism about safety practices at work.
8. Input your ideas into the safety planning and programmes.
9. Be careful to make sure that your computer/VDU is set up correctly and the chair is at the right setting.
10. Make sure that sharp objects, such as scissors, knives and so on, are stored away safely.
11. Make sure that any chemicals in the workplace are stored correctly.
12. If you feel tired or start making silly mistakes, take a break or change tasks.
13. When lifting objects, check for weight and use the correct safety techniques.
14. If any equipment, such as a photocopier or printer, jams or breaks down, let someone qualified fix it rather than try to make it work yourself.
15. Look out for trailing cables and ripped carpet and move or fix them.
16. Avoid stacking things too high on a desk or shelf, or storing items in corridors or walkways.
17. Avoid using alcohol, drugs or excessive medication (including headache pills or painkillers) during the day.
18. Know your workplace safety procedures and policies.
19. Know and be able to identify the hazards and risks in your workplace.
20. Know your fire safety procedures.

Notes

PLEASE COMPLETE BEFORE CONTINUING

What Causes Accidents?

Take time to reflect on the question below, making some notes in the space provided.
Have there been any accidents or injuries in your workplace (or in your experience). If so, what do you think might have caused them or been contributing factors?

INCIDENT	PROBABLE CAUSES

Please complete before continuing

Chapter 5
Learning Review

This chapter looks at practical ways of transferring key learning points to your own work situation.

Each exercise must be completed fully and reviewed with your manager, colleagues, safety representative or workmates.

Before starting this chapter, please take a few moments to make a note of any ideas or actions in the learning diary and log in Chapter 1.

TEST YOUR KNOWLEDGE (1)

1. List three points to remember when lifting or picking up items.

    1.

    2.

    3.

2. List three things you could check to do with electricity and electrical appliances.

    1.

    2.

    3.

3. List three causes of slips or trips in a workplace.

    1.

    2.

    3.

4. Explain exactly what you would do on hearing the fire alarm.

5. How often should safety assessments/inspections be carried out?

6. Who is qualified to carry these out?

7. If an accident takes place at work, what is the company's legal position?

PLEASE COMPLETE BEFORE CONTINUING

TEST YOUR KNOWLEDGE (2)

1. List three things that employers are legally bound to do:

    1.

    2.

    3.

2. List three things that employees are legally bound to do:

    1.

    2.

    3.

3. A Safety Inspector has right of entry, right to interview and take samples with or without an organization's permission.

    TRUE/FALSE?

4. List three pieces of recent EU legislation that may affect you in the workplace.

    1.

    2.

    3.

5. List three essentials that must be provided for in the workplace.

    1.

    2.

    3.

PLEASE COMPLETE BEFORE CONTINUING

TEST YOUR KNOWLEDGE (3)

1. List three hazards with the potential to cause harm in your workplace.

   1.

   2.

   3.

2. List three common causes of accidents in the workplace.

   1.

   2.

   3.

3. What is the difference between a hazard and a risk?

4. What does HASAWA stand for?

5. Small firms (employing less than 50 employees) have a worse record of safety than larger firms.

   TRUE/FALSE?

6. How many accidents occur at work every year?

   a) 600 000

   b) 1 600 000

   c) 2 200 000

7. How many people, at any one time, are suffering ill-health, either caused or made worse by work conditions?

   a) 600 000

   b) 1 600 000

   c) 2 200 000

PLEASE COMPLETE BEFORE CONTINUING

Test Your Knowledge (4)

1. How many working days are lost each year due to health and safety-related accidents, sickness or injury?

   a) 10 million

   b) 20 million

   c) 30 million

2. What is the most common form of accident/injury?

3. How many people are killed at work every year?

   a) 200

   b) 400

   c) 600

4. What are the most common accidents or injury in your organization? (You will need to research this separately.)

5. List three negative attitudes that contribute to accidents.

   1.

   2.

   3.

6. List three positive attitudes that prevent accident, illness and injury.

   1.

   2.

   3.

Please complete before continuing

CASE STUDY (1)

Read the following case study, and then answer the question on the following page.

Jim Walker was one of those types of people that you couldn't help liking. Nothing was ever too much trouble and he never had a bad word to say about anything. This made his foreman Dave's task even more difficult. Dave was concerned about Jim's lack of attention when it came to safety. He frequently and plainly disregarded instructions when it came to the wearing of safety equipment. For instance, he only wore safety gloves when his hands got cold and refused to wear the safety boots when it was too hot. The company had invested much more time and money over the past 12 months in new safety guards, protective equipment and notices, and yet the workplace didn't seem to be much safer for it.

Jim's pranks regarding using the warehouse loading bay for cricket practice also nearly got him, and everybody else, in hot water. A French juggernaut unexpectedly swung round the corner and it was only the air-brakes and the driver's quick reactions that avoided a nasty accident. The driver was furious and this nearly led to a scuffle as well. (What made it worse was that the driver didn't know what cricket was!)

However, Dave had now begun to notice that several other of the younger lads had picked up Jim's bad habits. Goggles, shoes and other protective equipment were not being worn, and sloppy behaviour was beginning to creep in. For instance, the warehouse was a mess, with boxes left in gangways and packing areas not properly laid out.

Dave knew that if he didn't do something soon, it would only be a matter of time before there was an accident or injury. He had tried talking to Jim about it, and Jim had listened and promised to be more careful, but nothing changed – at least not for more than a few days.

Dave was at a loss what to do. Should he get tough with Jim and risk upsetting him, and probably everybody else, or get tough with everybody in the hope that Jim would toe the line? What other way could he solve the problem other than through confrontation?

Case Study (1): Question

If Dave asked you for advice, what would you say he should do?

Please complete before continuing

CASE STUDY (2)

Read the following accident report and then answer the questions on the following page.

Accident Report

In the process of securing scaffolding in preparation for the rendering of a wall and repair of roof tiles, three poles and six clamps were dropped from a height of 30 feet when an inexperienced and unsupervised subcontractor slipped while securing them.

They caused over £2000 of damage to a company vehicle and, more seriously, injured three people, one of whom suffered serious head injuries. Altogether, the victims spent over eight weeks off work. However, because two of them were not wearing safety helmets at the time, they received no compensation and received written warnings, as this was in breach of company regulations.

Work was delayed by three days while the accident was investigated, which in turn caused further delay in other work. The cost of this disruption was considerable.

Case Study (2): Questions

Please consider the following questions, making your answers as clear and specific as you can.

1. *What do you think caused the accident?*

2. *Who was at fault?*

3. *What lessons can be learned?*

4. *What would you do now?*

Please complete before continuing

Appendix Suggested Answers to the Knowledge Tests

Test your knowledge (1): suggested answers

1. Bend with your legs
   Examine the load first
   Hold close to body
2. Wiring and trailing leads
   Fuse loading
   Worn cables
3. Trailing leads
   Ripped or damaged flooring
   Wet conditions
4. Leave the building immediately by the nearest exit, closing all doors and windows. Do not use lifts.
5. Every 12 months, or whenever conditions or work routines change significantly.
6. A 'competent' person.
7. The company must prove that it is not liable by demonstrating that all reasonable precautions had been taken.

Test your knowledge (2): suggested answers

1. Answers could include, but are not limited to, the following:

   Provide adequate training and supervision
   Provide PPE if needed
   Create safe working procedures
   Provide sufficient lighting
   Mark fire exits
   Display safety notices
   Keep the workplace tidy
   Provide adequate heating and ventilation
2. Answers could include, but are not limited to, the following:

   Follow safety rules
   Use all care and consideration
   Wear safety equipment that is provided
   Read safety notices
   Notify hazards
   Follow safe working practice
   Avoid working in a manner that might cause an accident or injury to self or others
3. TRUE
4. Answers could include, but are not limited to, the following:

   Manual Handling Operations Regulations 1992
   Display Screen Equipment Regulations 1992
   PPE Regulations 1992
   Provision and Use of Work Equipment Regulations 1992
   Workplace (Health, Safety and Welfare) Regulations 1992
   Safety Signs and Signals Regulations 1996
   Noise at Work Regulations 1989

5. Answers could include, but are not limited to, the following:

   Heat and ventilation
   Toilet and washing facilities
   Safety notices
   Fire exits
   Fire-fighting equipment
   A trained first-aider

Test your knowledge (3): suggested answers

1. Answers could include, but are not limited to, the following:

   Electric equipment
   Lifting operations
   Hazardous chemicals
   Working from heights
   Compressed air equipment
   Poorly maintained or serviced tools and equipment
   Fork-lift trucks

2. Answers could include, but are not limited to, the following:

   Slips, trips and falls
   Manual handling
   Working from heights – falls
   Untidy work area
   Carelessness
   Trailing cables

3. A hazard is a danger; a risk is the chance of that danger turning into an accident or injury.

4. Health and Safety at Work Act

5. TRUE

6. b) 1 600 000

7. c) 2 200 000

Test your knowledge (4): suggested answers

1. c) 30 million

2. Manual handling, followed by slips and trips

3. c) 600

4. Student's own response

5. Any three of the following:

   Overconfidence
   Laziness
   Stubbornness
   Impatience
   Ignorance
   Showing off
   Forgetfulness

6. Taking safety seriously
   Finding out information
   Spotting hazards and risks

Case studies

The purpose of these case study exercises is for students to apply safety knowledge and awareness.

Whilst there are no 'right' answers, students should highlight legal regulations and standards that have been broken and practical ways of enforcing these.

Lessons learnt from each situation should also be identified, both in terms of what caused the incidents and what could prevent them happening again in the future.

HEALTH AND SAFETY WORKBOOKS

The 10 workbooks in the series are:

Fire Safety	0 566 08059 1
Safety for Managers	0 566 08060 5
Personal Protective Equipment	0 566 08061 3
Safe Manual Handling	0 566 08062 1
Environmental Awareness	0 566 08063 X
Display Screen Equipment	0 566 08064 8
Hazardous Substances	0 566 08065 6
Risk Assessment	0 566 08066 4
Safety at Work	0 566 08067 2
Office Safety	0 566 08068 0

Complete sets of all 10 workbooks are available as are multiple copies of each single title. In each case, 10 titles or 10 copies (or multiples of the same) may be purchased for the price of eight.

Print or photocopy masters

A complete set of print or photocopy masters for all 10 workbooks is available with a licence for reproducing the materials for use within your organization.

Customized editions

Customized or badged editions of all 10 workbooks, tailored to the needs of your organization and the house-style of your learning resources, are also available.

For further details of complete sets, multiple copies, photocopy/print masters or customized editions please contact Richard Dowling in the Gower Customer Service Department on 01252 317700.